ROSEN Verified
NATURAL DISASTERS

BLIZZARDS

Beatrice Harris

ROSEN
PUBLISHING

Published in 2023 by The Rosen Publishing Group, Inc.
2544 Clinton Street, Buffalo, NY 14224

First Edition

Editor: Greg Roza
Book Design: Michael Flynn

Photo Credits: Cover DenisTangneyJr/iStock; (series background) PinkPueblo/Shutterstock.com; p. 5 Ipedan/Shutterstock.com; p. 7 justoomm/Shutterstock.com; p. 9 Andrei Stepanov/Shutterstock.com; p. 11 (bottom) Jefunne/Shutterstock.com; p. 11 (top) OSweetNature/Shutterstock.com; p. 12 Dmitrii Brodovoi/Shutterstock.com; p. 13 VectorMine/Shutterstock.com; p. 15 (Hawaii) Luc Kohnen/Shutterstock.com; p. 15 (Texas) Stacey Welu/Shutterstock.com; p. 17 (Doppler tower) Sergey Sivkov/Shutterstock.com; p. 17 (radar) https://commons.wikimedia.org/wiki/File:TDWR_and_NEXRAD_Refl_Compared_vert.png; p. 18 courtesy of US Navy; p. 19 Marik97/Shutterstock.com; p. 21 avtk/Shutterstock.com; p. 23 (top)Danny E Hooks/Shutterstock.com; p. 23 (bottom) CreativeAngela/Shutterstock.com; p. 24 https://commons.wikimedia.org/wiki/File:Wind_chill.png; p. 25 Suzannne Tucker/Shutterstock.com; p. 27 Pheelings media/Shutterstock.com; p. 29 https://commons.wikimedia.org/wiki/File:Brooklyn_Museum_-_Blizzard_of_March_1888,_Brooklyn_-_Breading_G._Way_-_overall.jpg; p. 30 https://commons.wikimedia.org/wiki/File:30_Jan_1977_Tonawanda,_NY_(Parker_Blvd_between_Brompton_Rd_and_Ellicott_Creek_Rd).jpg; p. 31 courtesy of NOAA; p. 32 Paul Stringer/Shutterstock.com; p. 33 Yeongsik Im/Shutterstock.com; p. 35 https://commons.wikimedia.org/wiki/File:Route-128S_Needham-MA_Blizzard-1978_a.jpg; p. 37 Krasula/Shutterstock.com; p. 38 Flipser/Shutterstock.com; p. 39 Indigo Fish/Shutterstock.com; p. 40 sockagphoto/Shutterstock.com; p. 41 DejaVuDesigns/Shutterstock.com; p. 43 Rajat Chmria/Shutterstock.com; p. 45 Roschetzky Photography/Shutterstock.com.

Cataloging-in-Publication Data

Names: Harris, Beatrice
Title: Blizzards, Beatrice Harris
Description: New York: Rosen Publishing, [2023] | Series: Rosen Verified : Natural Disasters | Includes index.
Identifiers: LCCN 2022028997 | ISBN: 9781499469585 (library binding) | 9781499469578 (paperback) | 9781499469592 (ebook)
Subjects: LCSH: Blizzards - Juvenile literature
Classifications: LCC QC926.37 .H38 2023 | DDC 551.55

Manufactured in the United States of America

CPSIA Compliance Information: Batch #CWRYA23. For further information contact Rosen Publishing at 1-800-237-9932.

CONTENTS

MORE THAN SNOW

Stormy weather can be really different. Some storms are loud and bright with thunder and lightning. Several kinds of **precipitation** may fall. Some storms don't have any precipitation, just winds. Storms happen in every season on every part of Earth. In winter, blizzards can strike.

What does the word "blizzard" make you picture? Do you think of a lot of snow blowing around? If so, you're on the right track. A blizzard is a dangerous snowstorm. Only snowstorms with the worst conditions are called blizzards. Blizzard winds blow so much snow around it's hard to see. It's unsafe to be outside for too long in such cold winds.

Blizzards can be natural disasters. Natural disasters are events in nature that cause great damage, or harm, and deaths. Blizzard snows

The word "blizzard" was first used to mean a bad snowstorm in 1870. It appeared in an Iowa newspaper.

and winds damage people's property and threaten their lives. They can block roads and ruin buildings. They can snap electricity, internet, and phone lines. They can cut people off from heat sources, food and water, and other people.

In this book, you'll learn how blizzards form and how scientists can **predict** them. Most important, you'll learn how to keep yourself and others safe from a blizzard's dangers.

WHAT IS A BLIZZARD?

The U.S. National Weather Service has a **definition** for a blizzard. Any storm that doesn't match all parts of this definition is a different kind of storm.

First, winds must blow at least 35 miles (56 km) per hour. Winds might be slower at times but **gust** to this speed or faster. Next, the storm must have snow. The snow may fall from clouds, but it doesn't have to. Winds may blow around snow that fell before the storm.

VERIFIED

Check out U.S. weather on the National Weather Service's website: **https://www.weather.gov/**

A blizzard that blows around snow that was already on the ground is called a ground blizzard.

The blowing snow creates the next part of a blizzard. **Visibility** is poor. That means it's hard to see around you through all the snow in the air. In a blizzard, you can't see farther than 1/4 mile (0.4 km). That makes it unsafe to drive. Even walking around can be dangerous. People lose their way.

Finally, all these conditions must last for at least three hours for a storm to be called a blizzard. Some blizzards last for days.

BLIZZARD CONDITIONS

- Large amount of falling or blowing snow
- Ongoing winds or gusts of at least 35 miles (56 km) per hour
- Visibility less than 1/4 mile (0.4 km)
- Three hours or longer

Notice how little you can see of the city in the background of this blizzard photo. This is called a whiteout.

HOW DO THEY HAPPEN?

All winter storms begin when two large air masses meet. One has cold, dry air. The other has warm, **moist** air. Often, the cold air mass moves down from the north, and the warm air mass moves up from the south. Where they meet is called a front.

Warm, moist air rises, and the water vapor in it forms clouds. If temperatures are cold in the **atmosphere**, the water vapor forms ice crystals. This ice combines to make snow.

FAST FACT

It must be at least 32°Fahrenheit (0°Celsius) in the atmosphere for snow to form.

HOW SNOWFLAKES FORM

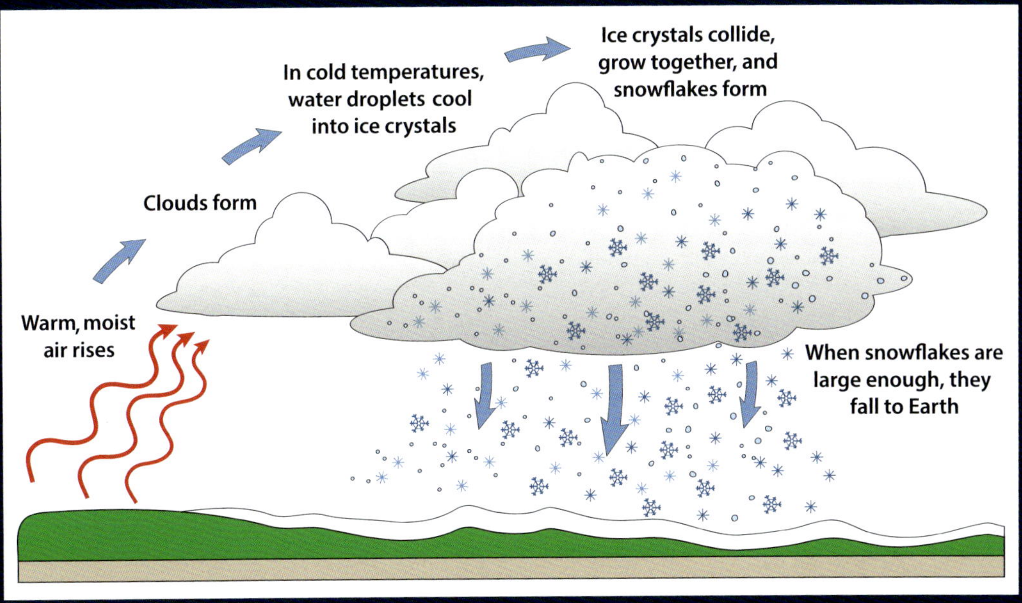

Snow may fall even if the temperature near the ground isn't 32°F (0°C). However, if it's too warm, rain may fall instead.

Cold and warm air masses don't mix. They push against each other. If the warm air mass moves in, it's called a warm front. It may bring light precipitation. If the cold air mass moves in, it's called a cold front. It may bring heavy precipitation.

Cold and warm air masses have different air pressure. Air pressure, or atmospheric pressure, is the measure of the weight of air above us. Air in high-pressure systems rushes into areas of low pressure. This creates wind.

FAST FACT

An air mass may move over warmer lakes or oceans and pick up water vapor. This can make even more rain or snow in the clouds. This is how events like lake-effect snows and nor'easters form.

VERIFIED
There's much more to learn about wind:
scijinks.gov/wind/

ATMOSPHERIC FRONTS

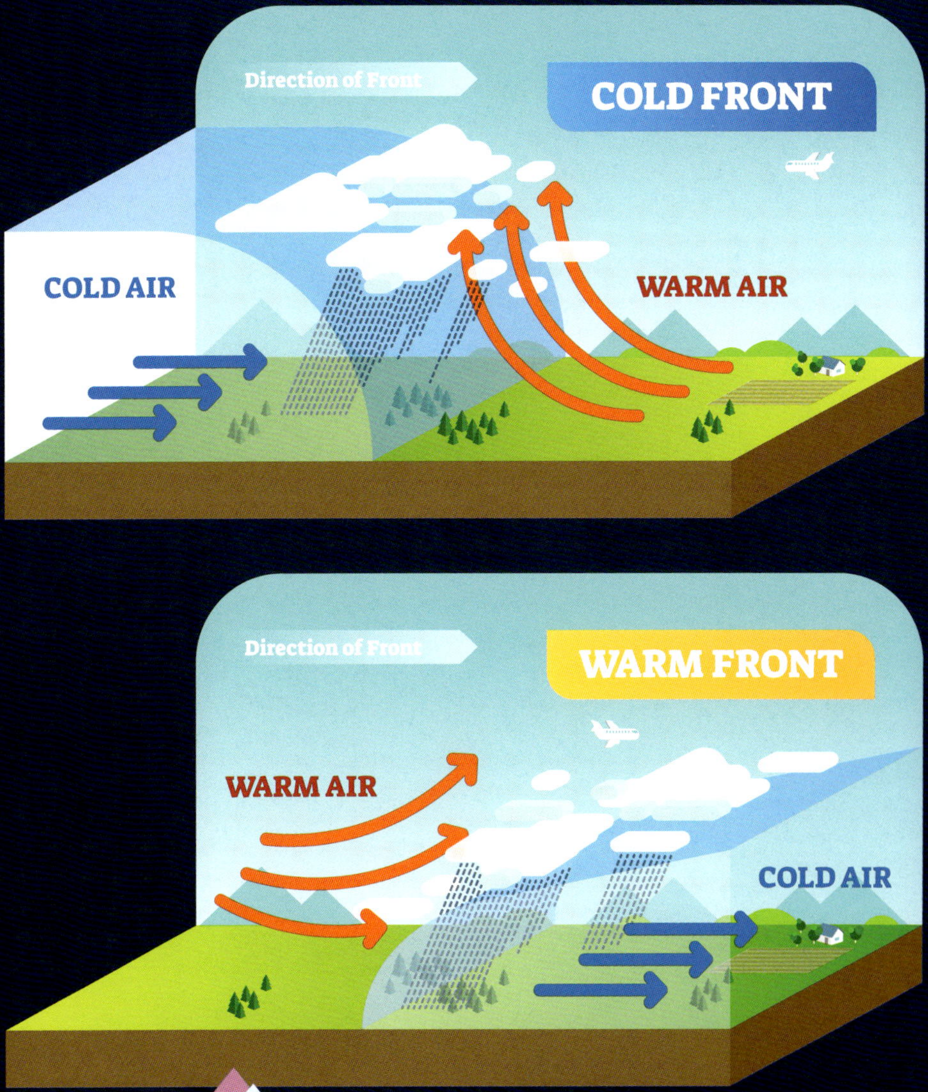

Blizzards require rising warm, wet air. Here are two kinds of weather fronts.

WHERE DO THEY HAPPEN?

In the United States, blizzards happen in the Great Plains, the upper Midwest, the Great Lakes area, and the Northeast. These places have cold winters. However, blizzards can occur as far south as Texas. In Canada, blizzards happen in the prairies of the west, eastern Ontario, and the eastern Arctic area.

Northern Europe, Northern Asia, and Antarctica also have blizzards. Blizzards can happen in warmer areas with mountains too. That's because air is cold at high **altitudes**.

FAST FACT

Hawaii's Big Island is home to high mountains. A blizzard formed in these mountains in 2021.

Snow can fall in places like San Antonio, Texas, but it's rare.

FORECASTING

Meteorology is the study of weather and the atmosphere. Scientists who study meteorology are called meteorologists. They use **technology** to forecast the weather. Forecasting means looking at present conditions to predict future weather.

Weather radar is a forecasting tool. It sends out radio waves. The waves bounce back when they hit objects, such as ice crystals. The radio waves give meteorologists information about storms that are forming.

Barometers are tools that measure air pressure. A sharp drop in air pressure can mean storms are approaching.

VERIFIED

Read more about how Doppler radar works:
www.weather.gov/iwx/wsr_88d

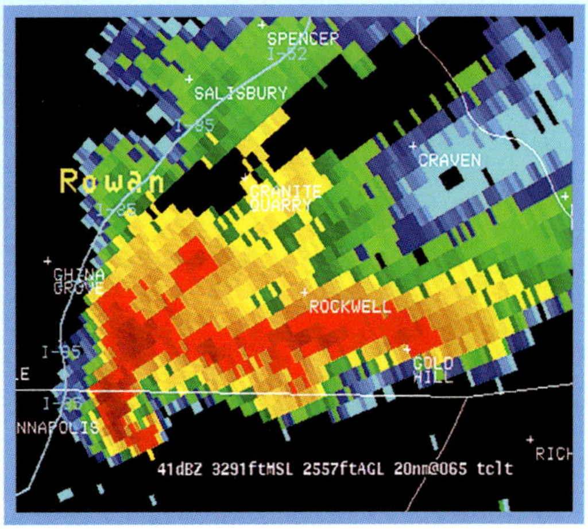

Doppler radar is a kind of weather radar. The radio waves that bounce back can tell scientists the velocity of a storm, or its speed and direction. A Doppler tower and image are shown here.

Weather satellites are machines that orbit, or circle, Earth. They send back storm images. Radiosondes are tools sent into the air with weather balloons. They collect **data** about temperature, air pressure, moisture in the air, and wind speed and direction.

VERIFIED

Check out more about radiosondes and how they're used:
www.weather.gov/jetstream/radiosondes

Computers have become a valuable tool for meteorologists.

Collected data is used with computer programs to figure out the forecast. Meteorologists report if conditions are right for a blizzard. They use the terms "blizzard watch" and "blizzard warning." A watch means a blizzard is possible. A warning means a blizzard is about to happen or is happening.

TERMS TO REMEMBER

BLIZZARD WATCH
a blizzard is possible

BLIZZARD WARNING
a blizzard is about to happen or is happening

SPREADING THE WORD

Meteorologists who work for TV stations **issue** blizzard watches and warnings on the news. You'll also hear about warnings and watches on the radio. You might read about them on websites and in newspapers too.

Some meteorologists work for the National Weather Service (NWS). When a blizzard is expected, the local NWS office lets officials and the public know. The NWS is part of the U.S. National Oceanic and Atmospheric Administration (NOAA). NOAA has 1,025 radio stations that provide up-to-date weather.

VERIFIED

Check out more about how NOAA and the NWS save lives: **www.noaa.gov/weather**

The NWS sends out **emergency** weather messages to people's smartphones.

WHY SO DEADLY?

You might wonder how a blizzard can be deadly. During blizzards, people die in car accidents. Falling trees and branches are also dangerous. Some people have heart attacks when shoveling snow. People without enough food and clean water are at risk too.

Carbon monoxide poisoning is also a risk. Snow can cover heating **vents**. Carbon monoxide put out by furnaces can't escape and comes into the home. Carbon monoxide can fill a car if the tailpipe is blocked by snow and the engine is running.

SYMPTOMS OF CARBON MONOXIDE POISONING

- headaches
- sleepiness
- **confusion**
- easily annoyed
- stomachache
- vision troubles

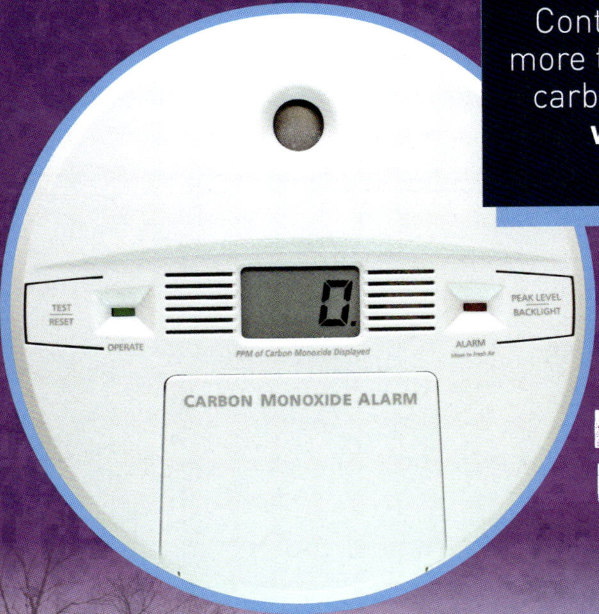

CARBON MONOXIDE DETECTOR

VERIFIED

The CDC (Centers for Disease Control and Prevention) has more to share about preventing carbon monoxide poisoning: **www.cdc.gov/dotw/carbonmonoxide/**

Other sources of carbon monoxide poisoning are gas **generators**, room heaters, and blocked chimneys. Make sure your home's carbon monoxide detectors are working throughout the year.

High winds and cold temperatures can also be deadly. People can get lost in whiteout conditions. Strong winds also produce low windchill temperatures.

Windchill is how cold a wind feels on human skin. Strong, cold winds can feel much colder than the air temperature. People who are outside in low windchill temperatures can get **frostbite**. Mild frostbite may not cause permanent, or lasting, damage to skin. But deep frostbite can permanently damage skin, muscle, and even bone.

WINDCHILL CHART

Wind (mph) \ Temperature (°F)	40	35	30	25	20	15	10	5	0	-5	-10	-15	-20	-25	-30	-35	-40	-45
Calm																		
5	36	31	25	19	13	7	1	-5	-11	-16	-22	-28	-34	-40	-46	-52	-57	-63
10	34	27	21	15	9	3	-4	-10	-16	-22	-28	-35	-41	-47	-53	-59	-66	-72
15	32	25	19	13	6	0	-7	-13	-19	-26	-32	-39	-45	-51	-58	-64	-71	-77
20	30	24	17	11	4	-2	-9	-15	-22	-29	-35	-42	-48	-55	-61	-68	-74	-81
25	29	23	16	9	3	-4	-11	-17	-24	-31	-37	-44	-51	-58	-64	-71	-78	-84
30	28	22	15	8	1	-5	-12	-19	-26	-33	-39	-46	-53	-60	-67	-73	-80	-87
35	28	21	14	7	0	-7	-14	-21	-27	-34	-41	-48	-55	-62	-69	-76	-82	-89
40	27	20	13	6	-1	-8	-15	-22	-29	-36	-43	-50	-57	-64	-71	-78	-84	-91
45	26	19	12	5	-2	-9	-16	-23	-30	-37	-44	-51	-58	-65	-72	-79	-86	-93
50	26	19	12	4	-3	-10	-17	-24	-31	-38	-45	-52	-60	-67	-74	-81	-88	-95
55	25	18	11	4	-3	-11	-18	-25	-32	-39	-46	-54	-61	-68	-75	-82	-89	-97
60	25	17	10	3	-4	-11	-19	-26	-33	-40	-48	-55	-62	-69	-76	-84	-91	-98

Frostbite Times: 30 minutes; 10 minutes; 5 minutes

$$\text{Wind Chill } (^\circ F) = 35.74 + 0.6215T - 35.75(V^{0.16}) + 0.4275T(V^{0.16})$$

Where, T= Air Temperature (°F) V= Wind Speed (mph)

The colors on this chart tell you how fast bare skin could get frostbite in windchill conditions.

Hypothermia is another condition to fear in blizzard weather. Hypothermia happens when a person's body temperature gets very low. The human body uses up energy creating its own heat. In cold conditions, it loses heat too fast. It can't create more heat quickly enough. Body temperature starts to drop from its normal temperature, around 98°F (37°C). If it gets below 95°F (35°C), hypothermia may occur.

Low body temperature is dangerous to human body systems. The heart and lungs may stop working. People with hypothermia need medical care.

WARNING SIGNS OF HYPOTHERMIA

- constant shivering
- memory loss
- feeling lost or confused
- being hard to understand
- sleepiness

Older people and babies are at most risk of hypothermia. It can even happen in a cold home.

HISTORIC BLIZZARDS

Certain blizzards throughout history show just how dangerous blizzards can be. The storm called the Great White Hurricane hit the East Coast of the United States in March 1888. Over 50 inches (1.3 m) of snow fell in three days in some places.

More than 400 people died, including about 200 in New York City alone. Over 100 people died at sea as blizzard winds created huge waves that sank ships. After the blizzard, some cities decided to build underground transportation systems.

FAST FACT

In January 1888, the Children's Blizzard hit the upper Midwest. Many children died while trying to walk home from school in the blizzard.

People stand next to piles of snow in Brooklyn, New York, after the Great White Hurricane of 1888.

The northern United States and Canada experience snowy weather more often in winter. But blizzards still stop everyday life there. The Blizzard of '77 in Western New York and Southern Ontario was one of the worst.

FAST FACT

The National Guard was called in to help clear snow from the Blizzard of '77. The National Guard is a military force that becomes part of the army and air force when needed by the federal government.

Between January 28 and February 1, 1977, wind gusts in Buffalo were as powerful as 69 miles (111 km) per hour. Windchill temperatures dropped to -59°F (-50.6°C). Only about 1 foot (30 cm) of snow fell. However, the strong winds whipped up the large amount of snow that had fallen before the storm.

After the Blizzard of '77, snowdrifts covered houses and cars. A snowdrift is a large snow pile created by wind. Emergency workers search for people trapped in a car in this photo.

The deadliest blizzard in history didn't happen in the United States. It struck Iran, a country in the Middle East. The five-day storm began February 3, 1972. It affected almost the entire nation.

As much as 24 feet (7.3 m) of snow fell in some parts of the country. So much fell that people were trapped in their homes. Some buildings were completely buried. People ran out of water and food. The storm cut off heat and power supplies. About 4,000 people died.

During a break in the wind and snow of the Iran Blizzard of 1972, rescue workers traveled to villages by helicopter. They dropped food hoping people could dig out and find it. Few did.

FORECASTS SAVE LIVES

In recent years, computer programs produce weather models that are much more **accurate** about blizzards than in the past. Forecasts are more likely to be correct farther ahead of time. They're also much more exact about the places that will be affected. This means people can be told ahead of time to stay home from school and work. They can avoid dangerous weather.

However, computer models of future storms are only as accurate as the data used. If one piece of information is missing, the whole forecast is off.

FAST FACT

A blizzard in the Northeast in 1978 was forecast to hit the morning of February 5. When it didn't, people thought the storm wasn't coming at all. They left their homes. The blizzard came hours later, trapping people in cars and at offices.

About 100 people died in the Blizzard of 1978. Thousands more were injured. Could a more accurate forecast have saved lives?

SNOW REMOVAL

Technology has helped us predict blizzards. Technology also helps us with the cleanup during and after snowfalls.

New York City has more than 500 snowplows and salt-spreading trucks. These trucks have **GPS**. An online map can tell people if and when a street was plowed.

Another kind of truck has a superhot snow melter. Snow is placed in a container and heated up. The water is poured into the sewer. This helps stop flooding that happens in some places when snow melts.

FAST FACT

Sometimes northern cities loan snowplows to southern cities hit by snowstorms.

Some trucks plow snow and spread salt. Salt lowers the freezing temperature of water, which is usually 32°F (0°C). This makes it less likely ice will form on roads.

AVOIDING DISASTER

Those who have been caught in dangerous situations during a blizzard have the same thought: they wish they had been prepared. As winter approaches, people can get ready for winter storms by keeping blankets, extra clothes, a shovel, and a full tank of gas in the car. It's best not to be on the road if a blizzard hits, but these things will help those who get stuck.

Call emergency services to get help on the road. Don't go out into a blizzard.

FAST FACT

If you're stuck in a car during a blizzard, the NWS suggests turning on the car for 10 minutes every hour to keep warm while waiting for help. Make sure the tailpipe isn't blocked.

People get stuck in their cars because of the amount of snow on the road or car accidents blocking the way.

If a blizzard is announced ahead of time, people can make sure they have enough food and water in their home. A first-aid kit is also needed in case of accidents. Blizzards can cause homes to lose power. A battery-powered radio can help keep people informed about what's going on. Flashlights are also a good idea.

Often when a blizzard hits, the local government bans travel. People should follow these orders. This helps keep roads clear for emergency vehicles and plows.

VERIFIED

Find out how the U.S. government wants you to be prepared for a blizzard:
www.ready.gov/winter-weather

Snow, ice, and wind can knock down trees and power lines. Homes can lose electricity and heat.

GLOBAL WARMING

Climate change is the long-term change in weather patterns on Earth. It's caused partly by human activities, such as burning oil and natural gas. These things are burned to create power and heat.

One effect of climate change is global warming. Earth's average temperature is increasing. Warmer air carries more moisture into the atmosphere. The temperature of the oceans is also rising, adding more moisture to the air. Scientists think this will create blizzards with even more powerful winds and larger amounts of snow.

VERIFIED

National Geographic explains more about how global warming happens:
education.nationalgeographic.org/resource/global-warming

Even the smallest temperature increase can cause a great difference on Earth, such as melting Arctic ice.

BE AWARE AND PREPARE

Some blizzards have been given powerful names, such as 2010's Snowmageddon. This mixes the words "snow" and "**Armageddon**." Other blizzards have been called Snowpocalypse (from "snow" and "**apocalypse**") and Snowzilla (from "snow" and "Godzilla"). These names might sound funny, but blizzards aren't funny. The names hint at how dangerous they can be.

Many scientists think we can expect shorter, warmer winters that will feature extreme weather events. On February 16, 2021, a winter storm covered about 73 percent of the United States in snow. This was the most snow ever recorded across the nation. The event brought blizzards to some places. Other places were hit with **tornadoes**. And still others faced deadly cold. The storm caused at least 246 deaths in Texas alone.

During the winter storm that hit the United States in February 2021, many people were without power, heat, and even water for days. This added to the death toll.

Technology makes it much easier to forecast blizzards, but no technology can stop them from forming. Knowing about blizzards can help people prepare for them, however. People can stay informed about weather. They can make sure their bodies are completely covered in windchill temperatures. They can keep their homes ready for emergencies. And once the wind and snow start blowing, they can stay home and stay safe.

GLOSSARY

accurate: Free of mistakes.

altitude: Height above sea level.

apocalypse: A great disaster.

Armageddon: A final destructive battle or conflict.

atmosphere: The mixture of gases that surround a planet.

carbon monoxide: A colorless, odorless, and poisonous gas that forms when carbon is not completely burned.

confusion: The state of being mixed up or unable to carry out tasks.

data: Information used to figure something out.

definition: A short statement explaining what a word or concept means.

emergency: An unexpected situation that needs quick action.

frostbite: A condition that can occur in which part of the body freezes or nearly freezes.

generator: A machine that uses movement to produce electricity.

GPS: Stands for Global Positioning System; a system that uses satellite signals to locate places.

gust: To blow in a sudden, powerful rush of wind; also, the sudden wind that blows.

issue: To tell people something in an official and public way.

moist: Having some water; a bit wet.

precipitation: Rain, snow, sleet, or hail.

predict: To guess what will happen in the future based on facts or knowledge.

technology: A method that uses science to solve problems and the tools used to solve those problems.

tornado: A narrow, funnel-shaped, swirling wind that can cause great destruction as it passes over land.

vent: A small opening that allows air in and out.

visibility: The ability to see.

INDEX